The World's Oceans

PACIFIC OCEAN

Heather Kissock and Megan Kopp

LIGHTBOX
openlightbox.com

LIGHTBOX

Go to
www.openlightbox.com
and enter this book's
unique code.

ACCESS CODE

LBXT6695

Lightbox is an all-inclusive digital solution for the teaching and learning of curriculum topics in an original, groundbreaking way. Lightbox is based on National Curriculum Standards.

OPTIMIZED FOR

- ✓ **TABLETS**
- ✓ **WHITEBOARDS**
- ✓ **COMPUTERS**
- ✓ **AND MUCH MORE!**

STANDARD FEATURES OF LIGHTBOX

AUDIO High-quality narration using text-to-speech system

VIDEOS Embedded high-definition video clips

ACTIVITIES Printable PDFs that can be emailed and graded

WEBLINKS Curated links to external, child-safe resources

SLIDESHOWS Pictorial overviews of key concepts

INTERACTIVE MAPS Interactive maps and aerial satellite imagery

QUIZZES Ten multiple choice questions that are automatically graded and emailed for teacher assessment

KEY WORDS Matching key concepts to their definitions

VIDEOS

WEBLINKS

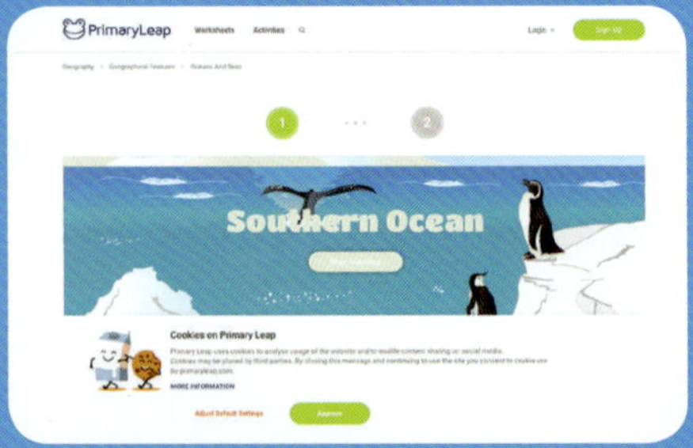

SLIDESHOWS

QUIZZES

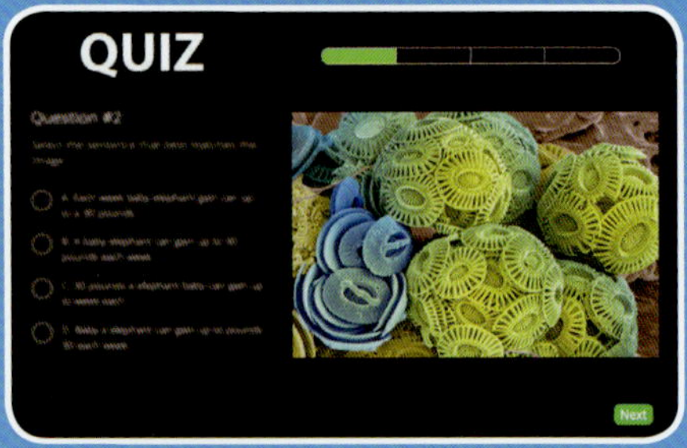

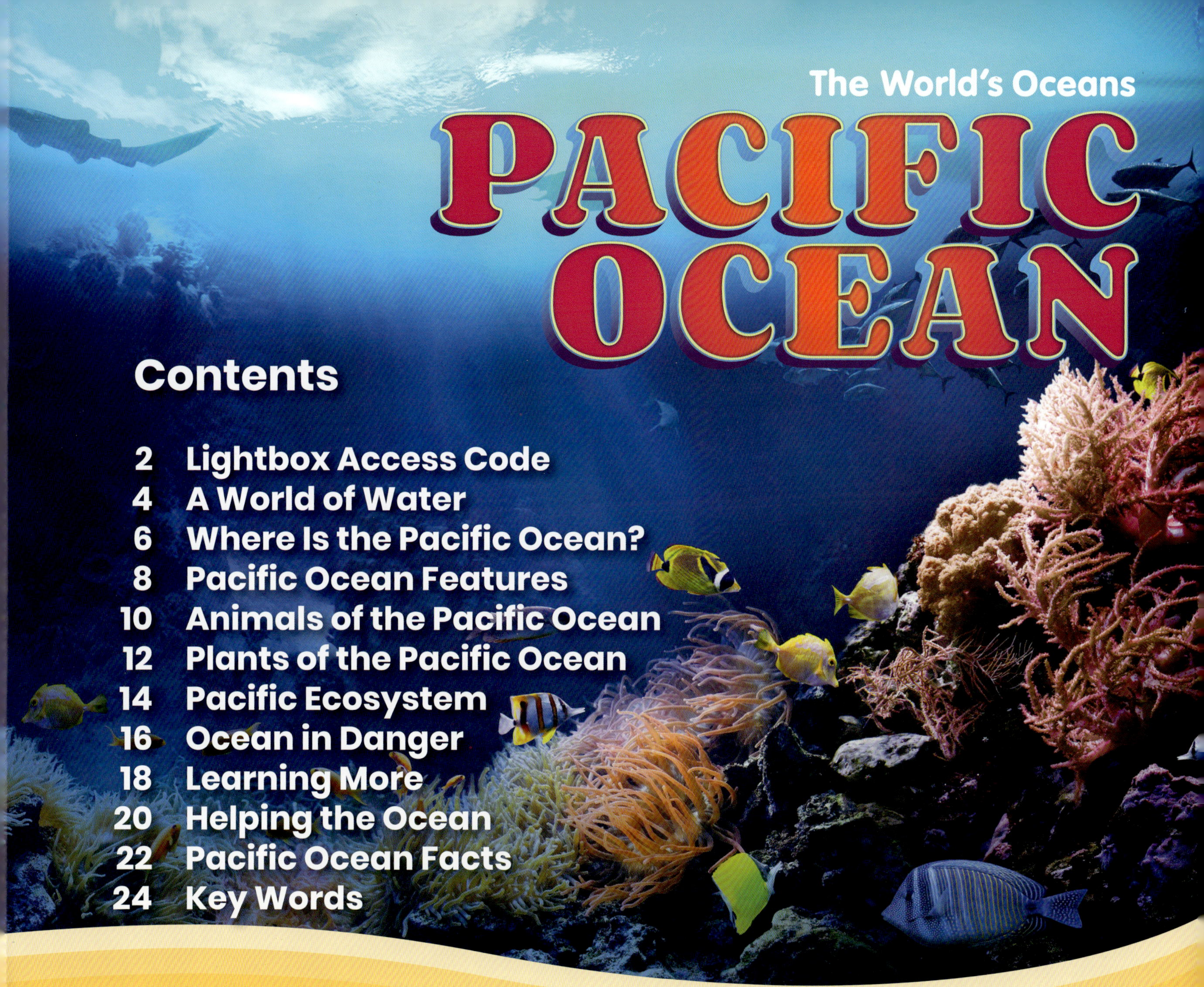

The World's Oceans

PACIFIC OCEAN

Contents

A WORLD OF WATER

Earth has five oceans. The Pacific Ocean is Earth's largest ocean.

The Pacific Ocean is so large that all 7 continents could fit inside it.

WHERE IS THE PACIFIC OCEAN?

The Pacific Ocean lies between the Arctic and Southern Oceans. It touches four continents.

THE WORLD'S OCEANS

PACIFIC OCEAN FEATURES

The Pacific Ocean has many coral reefs. Earth's largest coral reef system, the Great Barrier Reef, is in the Pacific Ocean.

The **Great Barrier Reef** is home to **more than 10,000 kinds** of **animals.**

ANIMALS OF THE PACIFIC OCEAN

A variety of animals can be found in and around the Pacific Ocean.

Common Murre

Sea Otter

Humpback Whale

Krill

Pacific Herring

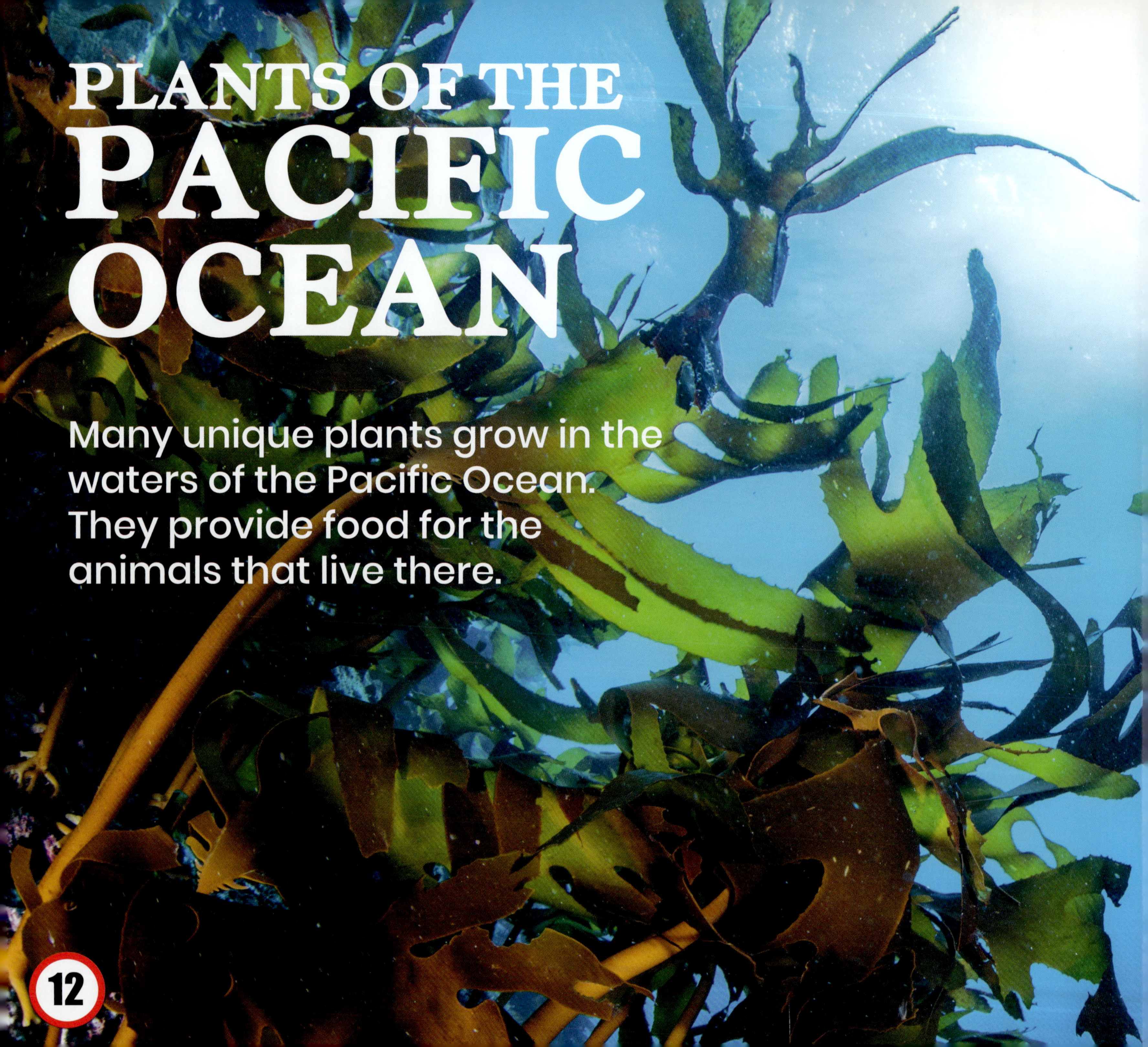

PLANTS OF THE PACIFIC OCEAN

Many unique plants grow in the waters of the Pacific Ocean. They provide food for the animals that live there.

Seaweed

Seagrass

Giant Kelp

Phytoplankton

PACIFIC ECOSYSTEM

The plants and animals that live in the Pacific Ocean need each other to survive. Together, they form an ecosystem.

PACIFIC FOOD CHAIN

Phytoplankton grow near the ocean's surface.

Phytoplankton are food for krill.

Krill are often eaten by Pacific herring.

Pacific herring are an important food source for humpback whales.

OCEAN IN DANGER

People are dumping their waste into the Pacific Ocean.

The animals and plants in the Pacific Ocean need clean waters to survive.

LEARNING MORE

Scientists and other people are studying the Pacific Ocean.

They want to know how they can make it clean again.

HELPING THE OCEAN

You can help the Pacific Ocean.

Buy a water bottle that you can use over and over again.

A reusable bottle helps keep plastics out of the ocean.

PACIFIC OCEAN FACTS

This page provides more detail about the interesting facts found in the book. Simply look at the corresponding page number to match the fact.

Pages 4–5

The Pacific Ocean covers almost one third of Earth's surface. It is about 5.9 million square miles (15.2 million square kilometers) in area, which makes it about 15 times the size of the United States. The Pacific Ocean's greatest known depth is 36,201 feet (11,034 meters).

Pages 6–7

The Pacific Ocean lies between Asia and Australia on the west, and North America and South America on the east. The ocean was named by Portuguese explorer Ferdinand Magellan. He called it *Mar Pacifico*, or "the peaceful sea." Magellan gave the ocean this name after he sailed it for three months without encountering any storms.

Pages 8–9

More than 75 percent of the world's coral reefs lie in the Pacific Ocean. Australia's Great Barrier Reef is home to about 400 coral species. Only living corals have color. Dead coral turns white. "Bleaching" is the term used when large amounts of coral die. This is often caused by a drastic change in the reef environment, such as a drop in temperature.

Pages 10–11

Many different animal species can be found in the Pacific Ocean. These include fish, mammals, and invertebrates. Animals in the Pacific have many adaptations that make them well suited to ocean life. To stay warm in the cold waters, sea otters rely on their thick fur. Gray whales use their blubber to stay both warm and buoyant.

Pages 12–13

Plant life in the Pacific Ocean can be found at various depths. Seagrasses grow along the ocean floor in shallow areas. Their deep roots help them survive the strong ocean currents. Large beds of seagrass have been known to even slow these currents.

Pages 14–15

All organisms need food, water, and space to survive. Producers, such as seagrasses, get their energy from the Sun. Consumers, such as crustaceans and fish, get energy by eating other organisms. This cycling of energy is called a food chain. Producers and consumers have an interdependent relationship. The space that they live in is called an ecosystem.

Pages 16–17

Marine debris is a growing hazard for Pacific wildlife. Animals can get caught in plastic wrapping, choke on inedible trash, or even be poisoned by chemicals found in the plastic. When marine wildlife such as fish ingest these chemicals, they become toxic to some degree. Humans then consume some of these toxic fish.

Pages 18–19

The Great Pacific Garbage Patch is a mass of marine debris floating in the North Pacific Ocean. According to studies, fish in this region eat up to 24,000 tons (22,000 metric tons) of plastic each year. The Ocean Conservancy is one of many groups working to create science-based solutions for a healthy ocean. Its Coastal Cleanup is the world's largest volunteer effort to clean up the ocean.

Pages 20–21

Items made of plastic create one of the biggest waste problems in the world today. Many items made of plastic are durable and meant to last a long time. This means that they do not decompose quickly. They stay in landfills and oceans for years. People can help keep plastic out of the ocean by being more cautious with their use and disposal of it.

KEY WORDS

Research has shown that as much as 65 percent of all written material published in English is made up of 300 words. These 300 words cannot be taught using pictures or learned by sounding them out. They must be recognized by sight. This book contains 63 common sight words to help young readers improve their reading fluency and comprehension. This book also teaches young readers several important content words, such as proper nouns. These words are paired with pictures to aid in learning and improve understanding.

Page	Sight Words First Appearance
4	a, Earth, has, is, of, the, water, world
5	all, could, it, large, so, that
6	and, between, four, where
8	great, in, many
9	animals, home, kinds, more, than, to
10	around, be, can, found
11	sea
12	food, for, grow, live, plants, there, they
14	an, each, need, other, together
15	are, by, important, near, often
16	into, people, their
19	again, how, know, make, want
20	help, you
21	keep, out, over, use

Page	Content Words First Appearance
4	oceans, Pacific Ocean
5	Africa, Antarctica, Asia, Australia, continents, Europe, North America, South America
6	Arctic Ocean, Southern Ocean
7	Atlantic Ocean, Indian Ocean
8	coral reefs, features, Great Barrier Reef, system
10	variety
11	common murre, humpback whale, krill, Pacific herring, sea otter
13	giant kelp, phytoplankton, seagrass, seaweed
14	ecosystem
15	food chain, source, surface
16	danger, waste
18	scientists
21	bottle, plastics

Published by Smartbook Media Inc.
14 Penn Plaza 9th Floor New York, NY 10122
Website: www.openlightbox.com

Library of Congress Cataloging-in-Publication Data
Names: Kissock, Heather, author. | Kopp, Megan, author.
Title: Pacific Ocean / Heather Kissock and Megan Kopp.
Description: New York, NY : Smartbook Media, Inc. 2021. | Series: The world's oceans | Audience: Ages 6-8 | Audience: Grades 2-3 |
Identifiers: LCCN 2020018615 (print) | LCCN 2020018616 (ebook) | ISBN 9781510555228 (library binding) | ISBN 9781510555235
Subjects: LCSH: Pacific Ocean--Juvenile literature. | Oceanography--Pacific Ocean--Juvenile literature. | Marine ecology--Pacific Ocean--Juvenile literature.
Classification: LCC GC771 .K57 2021 (print) | LCC GC771 (ebook) | DDC 551.46/15--dc23
LC record available at https://lccn.loc.gov/2020018615
LC ebook record available at https://lccn.loc.gov/2020018616

052020
110819

Printed in Guangzhou, China
1 2 3 4 5 6 7 8 9 0 24 23 22 21 20

Editor: Heather Kissock Designer: Terry Paulhus

Every reasonable effort has been made to trace ownership and to obtain permission to reprint copyright material. The publisher would be pleased to have any errors or omissions brought to its attention so that they may be corrected in subsequent printings. The publisher acknowledges Getty Images, Alamy, Shutterstock, iStock, Minden Pictures, and Wikimedia Commons as its primary image suppliers for this title.